ISSN No. 09750312

Vedic Science

International quarterly Research Journal of Indian Foundation for Vedic Science dedicated to the Vedic Sciences and Scientific Interpretation of Vedas and Allied Literature

Vol . 1

Editors

Dr. Ravi Prakash Arya
Dr. Surinder Kur

INDIAN FOUNDATION FOR VEDIC SCIENCE
H.O.1051, Sector-1, Rohtak, Haryana, India Ph. 01262-292580
Delhi Contact Ph. Nos.: 011-65188114; 09313033917
Emails: vedicscience@rediffmail.com; vedicscience@hotmail.com
Website : www.vedascience.com

First Edition

Kali era: 5100 (CE 1999)

Kalpa 1,97,29,49,100

Brahma era 15,55,21,97,29,49,100

ISSN No. ***09750312***

Disclaimer

The views expressed by authors in their papers are their own. It is not necessary for the publishers or editors to agree with them

Table of Contents

Objectives

Vedic Science Journal is published with the following objectives:

1). Towards preserving the Vedic scientific heritage globally.

2). To promote and promulgate the rational, secular and scientific outlook of the Vedic seers and scholiasts of ancient India.

3). To bring together the Vedic scholars, who have evinced a great interest for promotion of Vedic studies on scientific lines, and modern scientists or scholars interested into the Vedic scientific vision, both professionals and amateurs, irrespective of caste, creed, race, region or religions from India and other parts of the globe, to present the detailed accounts / results of their original findings, discoveries and researches into the Vedic sciences and scientific interpretation of the Vedas and allied literature.

4). To provide a forum or a platform to find out the solution to the various issues of mass and global concern in the light of eternal laws (social and natural) discovered by the ancient Seers of India.

5). To help the modern scientific knowledge synthesise with the Vedic one so that it may diffuse following the Vedic vision, for upkeep, advancement, safety and development of humankind and not for ecological crisis, destruction and extinction of living organisms.

6). To promote such a vision (as discovered by the seers of India in the times of yore) as could bring about social reforms leading to harmony, friendship, fraternity and common culture to the present day trouble torn world of humankind and help in the cause of national and international peace, and, thus, suggesting to the modern science for more and more constructive role and to the Vedic science for more and more active role for the upkeep and safety of living organisms and the planet.

7). It also attempts a thorough, dispassionate, and impartial scientific reappraisal of the extant Vedic and Allied literature.

8). This Journal will also try to formulate/ lay down various scientific theories and principles enshrined in the Vedas and Allied literature.

Editorial

Dear scholars / scientists! Vedic Science is an humble attempt :

Towards preserving the Vedic scientific heritage globally.

To promote and promulgate the rational and scientific out look of the Vedic seers and scholiasts of India.

To bring together the Vedic scholars who have evinced a great interest for promotion of Vedic studies on scientific lines and modern scientists or scholars interested into Vedic scientific vision, both professionals and amateurs irrespective of caste, creed , race, region or religion from India and all over the globe to present the detailed accounts / results of their original findings, discoveries and researches into Vedic sciences and scientific interpretation of the Vedas and allied literature.

To provide a forum or a platform to find out the solutions to the various issues of mass and global concern in the light of eternal laws (social and natural) discovered by the ancient Seers of India.

To help the modern scientific knowledge synthesize with the Vedic one so that it may diffuse following the Vedic vision for upkeep, advancement, safety and development of mankind and not for ecological crisis, destruction and extinction of mankind.

To promote such a vision (as discovered by the seers of this land in the times of yore) as could bring about social reforms leading to harmony, friendship, fraternity and common culture to the present day trouble torn world of mankind and help in the cause of national and international peace thus suggesting the modern science for more and more constructive role and Vedic science for more and more active role for the upkeep and safety of living organisms and the planet.

It also attempts a thorough, dispassionate and impartial scientific reappraisal of the extant Vedic and Allied literature. This journal will also try to lay down various scientific theories and principles enshrined in the Vedas and Allied literature.

The Scholars / Scientists who are keenly and actively engaged in the Vedic studies and research are hereby cordially invited to present the detailed accounts / results of their original findings, discoveries and researches into Vedic and allied literature by way of a research paper to the Vedic Science.

During this period several scholars have raised queries as to what is Vedic Science and how does and to what extent the Vedic Science differ from the modern science. All such queries will duly be answered in the next editorial.

Rediscovering Vedic Sarasvatī

Dr. Ravi Prakash Arya

Sarasvatī is the perennial river of Bhārata. It is the very soul of Indian culture and thickly embedded in the entire fabric of Indian psyche. In fact Sarasvatī and Indian culture both are interwoven. The origin of this great perennial culture of the world known as Vedic culture owes its origin to the very origin of Sarasvatī. But the question of the place of the origin of Sarasvatī is perplexing and vexing the scholars since the last quarter of 19th century. Scholars cominge from various disciplines have had their say as to the various origins of Sarasvatī basing their arguments on various evidences ranging from Sanskrit literature to the LANDSAT imagery and other modest techniques of research.

However most of the studies carried out so far are based either on the interpretation of LANDSAT imagery or field verifications in the Indian part of the desert or the archaeological remains associated with old river courses. Literary sources have though been taken into account by many scholars, but they are mainly confined to describing the greatness and various epithets of the mighty Sarasvatī or at the most they try to find some literary evidences in support of their postulations and conjectures formulated on the basis of findings from archaeological excavations and LANDSAT imagery regarding the origin of Sarasvatī. For instance, the evidences from archaeological remains associated with old river courses have suggested that a major river stemming mainly from the same source as the present Sutlej flowed through northern Rajsthan, Bahawalpur and Sind to the south-east of the present course of the Sutlej and Indus in the third millennium BC (Allchin Goudie and Hegde, 1987). It is also suggested that the river used to flow through Raini and Wahind rivers and then through eastern Nara and Hakra in the Sind province of Pakistan to the Rann of Kutch (Oldham 1893; Wadia 1938; Ali 1942; Stein 1942; Krishan 1960; Wilhelmy, 1969). The evidences gathered from the interpretation of aerial photographs and LANDSAT imagery tend to prove that the Sarasvatī used formerly to flow through a more easterly course in the Rajsthan part of the desert when the Luni river was one of its tributaries. Subsequently, the river shifted its course to the west several times and finally severed its link with the Luni (Ghose et al, 1978). The evidences from the aforementioned sources further lead to prove that when the river severed its link from Luni and shifted westward, it didn't occupy the Raini-Wahind-Nara- Hakra course in Pakistan.

But the main problem is that none of the scholars have so far been able to direct his archaeological or other studies in the light of numerous and varied information lying scattered in the vast Vedic and post-Vedic literature on the various phases of existence of Sarasvatī. In fact the vast and extensive informative data on various phases of existence of Sarasvatī has mixed up in such a manner that it becomes very difficult for a lay scholar, who is not well-acquainted with the Vedic dicta to systematically arrange and consolidate the data phase-wise and thereby to present an integrated account of the origin and flow of Sarasvatī during various phases ranging from Vedic down to the age of *Mahābhārata* (c.3152 BC.) and afterwards.

1.1 Origin of Sarasvatī in the Vedic Phase

The first origin of Sarasvatī in the Vedic phase goes with the very origin of waters on the earth and as such Sarasvatī is described in the Vedas as owing its origin to clouds. The Ṛgvedic seer Vasiṣṭha has clearly indicated in his observations that celestial *Parvatas* were the source of origin of Sarasvatī

ā no divo bṛhataḥ parvatādā
sarasvatī yajatāgantu yajñam
havaṁ devi jujuṣāṇā ghṛtācī
śagmo no vācamuśatī śṛṇout

From the great celestial *Parvata* Sarasvatī flow towards the *yañīya* site. It hears the voice of a person who offers the oblation of *Ghṛta* or clarified butter with the desire to receive it.'

Here the use of the phrase *divaḥ Parvata* clearly suggests that the celestial *parvata* here is not the mountain but the cloud in the sky.

The fact is that *parvata* or its synonyms have never been used for mountains in the earlier Vedic language. They always signified clouds. Yāska (*Nirukta* 1.20) observed in this connection as follows:

Giriṣṭhā Giristhāyī Giriḥ Parvataḥ samudgīṛṇo bhavatiParvavān Parvataḥ parva punaḥ pṛṇāteḥ prīṇātaṁ vārdhmāsaParva devānasmin prīṇantīti ut prakṛtītarat sandhiSāmānyāt meghasthāyī megho'pi Giriretasmādeva

Giri is known as *Parvata*, since it is situated at the higher place than the surface of the earth. *Parvata* is so called, as it is constituted of fragments. It is so called as it brings up the living beings on the earth with water, etc. These qualities are discernible in the clouds; hence a cloud is called *giri.*

The author of the present lines has already established at several places that Vedic dicta got formed in the background of cultural, geographical or geological or scientific scenario prevalent by then. The Indian culture is not 3,4 or10 thousand years old as is often considered. In fact this great perennial culture, as it is known from its chronology available in the *Purāṇas* and its era retained in the *Saṁkalpa pāṭ has*, is older than the oldest Himalayas. This culture flourished around 197 crore years ago when there were no mountains visible on the earth. So it was natural that Vedic people had no mountains to denote them by the term *Parvata.* In the light of the above discussion the oft-repeated Ṛgvedic *Mantra* revealed to the Vasiṣṭha.

ekā cetat sarasvatī nadinām śuciryatī

Giribhya ā samudrāt. (7.95.2)

cannot be taken for Sarasvatī as flowing from mountains to sea, rather it would be taken to mean Sarasvatī flowing from clouds to sea.

1.2 Brahmanic Origin

Second origin of Sarasvatī belongs to the *Brāhmaṇika* phase, during which it was known to have originated from *Plakṣa Prasravaṇa.* This fact has also been corroborated by several Puranic traditions also. *Plakṣa Prasravaṇa* of Brahmanic period point out oozing or thawing out of Sarasvatī from Himalayas as glaciated region. There are clear cut indications in the tradition of *Skand Purāṇa* (7.30.41) that Sarasvatī having approached the glaciers flowed on the earth through *Plakṣa.* The approach of Sarasvatī to glaciers was its precipitation in the form of snow flakes from clouds and its flow in the form of waters on the earth has been assigned to place called *Plakṣa.* In fact *Plakṣa* may be recalled as the place of origin of Sarasvatī during the period when the present day Himalayan region witnessed the first ever glaciation since the origin of earth.

Origin of Sarasvatī from *Plakṣa* has also been registered in the memories of other *Purāṇas* too, e.g. according to *Vāmana Purāṇa* (3-4), Perennial Sarasvatī originated from *Plakṣa* tree.

Plakṣa vṛkṣāt samudbhūtā Saricchreṣṭhā sanātanī

In *Bṛhannāradīya Purāṇa* (64-17) also Sarasvatī has been described as *Plakṣa Jātā*, i.e. originated from *Plakṣa.*

mārkaṇḍeyena muninā saṁtaptaṁ paramaṁ tapaḥ
tatra- tatra samāyātā Plakṣajātā Sarasvatī

In *Tāṇḍya Brāhmaṇa* (25-10-16) the distance between the place of drying up (Vinaśana) and point of rising of Sarasvatī has been measured as 44 Āśvins, the distance between Plakṣa and Vinaśana is as long as 44 Āśvins.

Catuścatvāriṁśadāśvināni sarasvatyā vinaśanāt Plakṣaḥ Prāsravaṇaḥ

Now the main question is as to what was the *Plakṣa* whether it was a name if place or a tree as has been remembered in several books. If it was a name of place, the question is where and how to locate it?

Plakṣa, in fact, was the name of a place, perhaps conspicuous with the presence of *Plakṣa* trees, it was called *Plakṣa.* So one should not get oneself confused with *Plakṣa* as tree or place. *Plakṣa* as tree is often known in vernaculars as Pilakhua or a fig tree. It is known in botanical terms as ficus religiosa. Due to rising of Sarasvatī from a place of *Plakṣa* trees. *Plakṣa* was attached a religious fervor and the tree is considered to be holy. Such a place, which was conspicuous with *Plakṣa* Tree is Pilakhua in western UP. This place is located near Hapur and Garhgaṅgā. Pilakhua is obviously named after Pilakhana, a corrupt form of *Plakṣa.* Now the major problem is to establish the location of *Plakṣa* as modern Pilakhua as per literary records of Vedas, *Purāṇas* or Epics. (1) In this regard a reference of *Vāmana Purāṇa* (3-4) may be cited. Accordingly Sarasvatī after crossing thousands of stones entered the Dvaitavana.

saiṣā śailasahasrāṇI vidārya ca mahā nadī
Praviṣṭā puṇyatoyaiṣā vanaṁ daivatamiti śrutam

N.L Day has located the area of Dvaitavana near about Deoband in modern UP which is 50 miles north of Meerut or ancient *Marudhanvan* (*Āraṇyaka Parva* 243.21:295.8:296.40 BORI, Poona)

(2) King Yudhiṣṭhira started from Kurujāṅgala towards the north on the bank of river Sarasvatī in search of some thick forest full of roots and fruits after sometime he entered Dvaitavana.

There is also a reference of Pāṇḍvas proceeding towards Sarasvatī to the north of Marudhanvan (Modern Meerut) Thus it is clear from the above cited references that route of Sarasvatī was once via or close to the area of Deoband located to the north of Marudhanvan (Meerut)

(3) Next is the location of Kāmyaka vana. According to Chapter 244 of Āryaṇyaka Parva, Kāmyaka vana was there almost in Marudeśa. here it may be clarified that Marudeśa i.e. Marusthal or desert land was not far from Hastinapur. From the north of Kāmyakavana began the forests of Himalayas. Thus the location of Dvaitavana on the banks of Sarasvatī and closeness of Dvaitavana and Kāmyakavana to the Himalayan hills and forests respectively reveals that Sarasvatī in Brahmanic period flowed through UP and *Plakṣa* was nothing but the modern Pilakhua. As per the author of *Geography of Mahābhārata* (P.213), from Plakṣa Prasravaṇa; Iśānudhyāsī Tīrtha was at a distance of about half a mile, i.e.. The name of Tīrtha is indicative of some place of worship of Śiva. It can also be located in or around Pilakhua.

(4) *Mahābhārata* (*Vanaparva*: 8.3-84) enumerates *Plakṣa* in the list of 113 Tīrthas lying in Kuru region. Kuru region is identified with modern regions of Thaneshwar, Delhi, Meerut and Bijnor between Sarasvatī and Gaṅgā rivers having its capital at Hastinapur (near modern Bijnor). See also *Geographical Horizon of Mahābhārata* P.137. *Jain Ādi Purāṇa* also locates Kurujāṅgala in between Sarasvatī and Gaṅgā (16.193).

Most striking features of Tīrthas enumerated in *Vanaparva* of *Mahābhārata* is that Gaṅgā-Sarasvatī Saṅgam (84.38) has also been enumerated as one of the Tīrthas of Kuru regions. This makes it explicitly clear that Sarasvatī and Gaṅgā must be joining each other at some place in Kuru region and meeting place of Gaṅgā and Sarasvatī in the Kuru region can be none other but the area around Garhgaṅgā and Pilakhua in western UP.

5) While recounting *Plakṣa* as the Tīrtha Of Kurukshetra, it is told that Sarasvatī originated from a Bamboo near Saugandhika Vana which has also been identified near Mandākinī river of Garhwal hills of UP. Here it may not be odd to relate a folk tale of Bhutias of Mana village lying in the Garhwal region of UP. The folk tradition of the villagers remembers the origin of Sarasvatī in the region of Mana pass at an altitude of 18,400 feet above sea level. This pass is surrounded by high peaks covered with huge mass of snow and ice and lies on the old trade route along Indo Tibetan border to Garhwal region of UP. From Mana Pass to Mana village, Sarasvatī covers a distance of 45 Kms. In the high altitude regions of the Himalayas. The altitude varies from 11,000 feet at Keshava Prayāg (the Meeting Point of Sarasvatī and Alaknandā) to more than 18,000 feet at the origin. As per folk tale the river goes down Pātāl Loka from Bhima Bridge, living a small stream to meet Alaknandā. The rest of river reappears at Prayāg rāj to form the Triveṇī. The fact is that before its junction with Alaknandā at Keshav Prayāg, the Sarasvatī falls in a deep and narrow gorge below Bhima Bridge, which is a natural rock bridge reportedly built by one of the five famous Pāṇḍva brothers Bhima during 12 years long wandering period. The gorge is 150 to 200 feet deep. The river Sarasvatī emerges from gorge after its sound and fury in white caps and meets Alaknandā to form Keshav Prayāg.

Thus from local tradition of folk tales also it is crystal clear that *Plakṣa* was a place below glaciated region and during Brahmanic period most probably it was Pilakhua situated on the route of Gaṅgā.

6) As pointed out above, according to *Tāṇḍya Brāhmaṇa* (Benaras 1936,25.10 .1-23) the **distance** from Vinaśana to Plakṣa Prasravaṇa is 44 *Āśvins*. At the same distance from *Plakṣa* is situated *Svarga loka*. So people who want to go to *Svarga loka* take a journey alongwith the stream of Sarasvatī.

Catuścatvāṁśadāśvināni sarasvatyā vinaśanāt Plakaḥ prāsravaṇaḥ tāvaditaḥ savargo lokaḥ sarasvatī sammitena adhvanā savargaṁ lokaṁ yanti

Here one thing must be known to the scholars and others that *Svarga loka* in Sanskrit language implies to the northern region or Garhwal region of Mandākinī river in upper Himalayan region. Hence as per reference of *Tāṇḍya Brāhmaṇa* cited above *Plakṣa* was situated at equi-distance from Vinaśana and Garhwal Himalayan region. And the sacrificers used to go to *Svarga loka* alongwith the stream of Sarasvatī. Here it may be clarified that *Plakṣa* was not located on the hilly terrain, rather it was located in plains. Sarasvatī's track from Garhwal region down to *Plakṣ a*, clearly points out that Pilakhua can only be the exact location.

Here it may also be given to understand that in Brahmanic period Sarasvatī never terminated at western sea, rather it terminated at Vinaśana itself around 44 *Āśvins* from *Plakṣa*. This fact also points out that in Brahmanic period Sarasvatī may not have been a big river as was the case in later period. In fact the Brahmanic period may be considered as the glaciated period. As per modern Geologists in

Triassic period i.e. 22 crore years ago northern hemisphere consisting of the present day Europe was still a desert, but the southern hemisphere consisting of Indian continent was covered with snow. The memory of this type of glaciation still exits in the memory of the *Purāṇas.* Thus during this Brahmanic phase only Sarasvatī had a small track to flow.

Sarasvatī's flow till western sea is described in the *Skand Purāṇa* in the context of draining the *Vaḍvānala* (7.33.40-41) to the western sea at Prabhāsa Pāṭan. *Vaḍvāgni* was nothing else but the submarine volcanic activities of Mānas sara or the Tethy's sea It is well known that Himalayan ranges of mountains formed from volcanoes caused by plate-tectonic upheavals at the submarine level of Mānas sea. Due to volcanic activities at the submarine level of Mānas sara, the upliftment of the bottom began to take place. This upliftment caused the dewatering of the sea. This outflow of uplifted Mānas sara was drained by Sarasvatī towards western sea at Prabhāsa Pāṭan. This period goes back to the age of the first phase of formation of the Himalayan ranges. Himalayan ranges began to form in Eocene period i.e. around 7 crore years ago, but according to Indian tradition the Himalayan ranges began to form in the beginning of *Vaivasvata Manvantara.* i.e. 12 crore years ago.

We also meet with such references in the *Purāṇas* as describe the origin of Sarasvatī form upper Himalayan ranges of mountains. According to the tradition of *Matsya Purāṇa* (121.64-65), Sarasvatī and Jyotiṣmatī rivers flowing towards western and eastern seas respectively issue from *Sarpa* or *Nāga* lake situated at the back of Hemakuṭa.

Paraspreṇa dviguṇā dharmataḥ kāmato'rthataḥ
Hemakuṭasya pṛṣṭhe tu sarpāṇāṁ tatsaraḥ smṛtaḥ—64
Sarasvatī prabhavati tasmājjyotiṣmati tu yā
Avaghāḍhe hyubhayataḥ samudrau pūrvapaścimau.—65

Vāyu Purāṇa (*Gaṅgā Avtāra-Varṇana*-Chp. 47) depicts the same fact only with a few miner variations, e.g. *Sarpa* or *Nāga* lake has been mentioned as Śyanā lake and Sarasvatī has been mentioned as Manasvinī.

In the reference given above, Hemakuṭa is modern Kailāśa range (See S.M. Ali, *Geography of Purāṇas*, People's publishing House, Delhi, 1973) and the Sarpa lake or Śyanā lake at the back of Kailash, according to Ali, is the lake Nak Tso which with the Pangong Tso forms an extensive water sheet which join the Shyok at its south eastern bend and then combined waters flow towards the western sea (Slim Ali, P.70-71)

Today also we find two river systems Sindhu and Brahmaputra issuing from Mānasarovar lakes i.e. Sindhu from Mānasarovar and Brahmaputra from Rakṣasa Tāla, whereby Sindhu enters into western sea and Brahmaputra in eastern sea. *Matsya Purāṇa* names Sindhu as Manasvinī due to its origin from Mānasarovar. *Vāyu Purāṇa* retains its old name Sarasvatī only. On the other hand Brahmaputra has been named as Jyotiṣmatī by both since it flows towards east or the direction of Jyotiṣpur or Prāg-Jyotiṣpur(Assam) i.e. the place of the first rising of sun in India.

In fact, Sindhu and Brahmaputra are the two separated streams of one and same Sarasvatī. They are the creation of Himalayan ranges. In the *Nadī Sūkta* of *Ṛgveda* Sindhu and Brhmaputra have not been mentioned, since the Ṛgvedic *Sūkta* was composed before the origin of Himalayan ranges. It was after the formation of the upper Himalayan region that the mighty flow of Sarasvatī got separated into two streams, well known in the tradition of *Padma Purāṇa* (5.18; 117-28; 123) and *Skanda Purāṇa* (7.35.26) as prācī Sarasvatī and Paścimābhimukhi Sarasvatī which are Sindhu and Brahmaputra respectively. *Mahābhārata Anuśāsana Parva* (134.15) clearly mentions that Sarasvatī originated first of all the rivers:

eṣā Sarasvatī puṇyā nadināmuttamā nadi
prathmā sarvasaritāṁ nadi sāgar gāminī

The above arguments and the present *śloka* of *Mahābhārata* clearly shows that Sarasvatī was the oldest stream, which was later known as the divided stream of the Sindhu and Brahmaputra.

Thus the origin of Sarasvatī after formation of upper Himalayan ranges as a divided stream from lakes variously called in the *Purāṇas* as Mānas lake or Sarpa lake or Nāga lake, was known as *Nāgodbheda*. *Nāk* literally means *Svarga*. i.e. upper Himalayan region.

Śivodbheda

In the last Himalayan phase, we find the origin of Siwalik hills. In fact, Kailash region was famous as the abode of Śiva (while in the glaciation period it was famous as a *Himvat* region). Later on when the hilly terrain developed at the lower regions of Kailash, that came to be known as flocks of Śiva or *Śivālaka* or Siwalik in the present sense of terms. Siwalik ranges are also known as *Himvatpāda* being the last phase of Himalayas.

With the origin of Siwalik ranges, we find a shift in connection of Sarasvatī from Sindhu or Brahmaputra (Jyotiṣmatī) to Gaṅgā. Its origin also came to be known as Śivodbheda like that of Gaṅgā's fall on the flocks of Śiva from *Svarga*.

Tradition of *Brahma Purāṇa* reflects an ample good light on this fact by way of narrating a story that Sarasvatī owing to the fear of Brahmā's growing love towards her submerged itself into the basin of Gaṅgā. In fact, the eastern region was always known as the *Brāhma* region or *Brahmarṣi deśa* or Burma as is well known today. And northern region as *Śaiva* region. Sarasvatī's fear of growing love of Brahmā clearly indicates the drainage of maximum waters of Sarasvatī towards east through its eastern stream/ channel called Brahmaputra. With the partition of Sarasvatī into Sindhu and Brahmaputra, its downward main flow got abandoned. This fact is revealed in the *Purāṇas* in a figurative manner. The *Purāṇa* narrates the story of

Sarasvatī's merging into the basin of Gaṅgā. Tradition of *Brahmavaivarta Purāṇa* also holds the view that Sarasvatī and Gaṅgā had the same origin. The one and the sameness of Gaṅgā and Sarasvatī is also substantiated by a tradition of *Vāmana Purāṇa* (Chap.42) wherein it is ruled that if one takes a dip into eastern flow of Sarasvatī he gets the benefit of taking a dip into Gaṅgā.

Pūrva pravāhe yaḥ snāti gaṅgāsnānaphalaṁ labhet

In fact, in post Siwalik phase eastern flow of Sarasvatī was associated with Gaṅgā only.

The connection between Sarasvatī and Gaṅgā could also be established on the basis of *Plakṣa* (or Pilakhua) as the origin of Sarasvatī during Brahmanic period. Ancient *Plakṣa* or modern Pilakhua being situated at a point close to the basin of Gaṅgā or Garhgaṅgā. Thus the original source of Vedic Sarasvatī (*Plakṣa*) and modern Gaṅgā (Garhgaṅgā) i.e. home of Gaṅgā being close to each other, the origin of Siwalik Sarasvatī, *Śivodbheda* Sarasvatī or lost Sarasvatī and Gaṅgā was recognized as one and same.

Delinking of Sarasvatī from Gaṅgā

The Puranic traditions further shed light on the delinking of Sarasvatī from Gaṅgā. *Brahmavaivarta Purāṇa* (Chp. 6.70-71) holds that there arose a dispute between Gaṅgā and Sarasvatī in which both exchange curses leading to a pledge by Gaṅgā to end Sarasvatī's existence. *Purāṇa* tells us that cursed by Gaṅgā to turn black, Sarasvatī descended on the holy place of Bhārata, which undoubtedly is Kurukshetra.

puṇyakṣetre hyājagāma bhārate sā Sarasvatī
gaṅgāśāpena kalayā svayaṁ tasthau hareḥ padam.

(*Brahmavaivarta Purāṇa,* 2.7.1)

Here Gaṅgā's curse to blacken Sarasvatī clearly reveals the fact that after severing its link with Gaṅgā, Sarasvatī's origin was associated with the place of origin of Yamunā whose waters are considered to be of black hue as compared to those of Gaṅgā.

This fact has also been described by the tradition of *Skanda Purāṇa* (7.33.40-41).As per the tradition the *Plakṣa* descended on the famous Yamunā tīrtha which was also known as the doorstep to the *Nāk* or heaven, i.e. upper Himalayas.

etad vai plakṣāvataraṇaṁ yamunātīrthamuttamaṁ
etad vai nākpṛṣṭhasya dvāramāhurmanīṣiṇaḥ

Yamunā also used to flow to the westward sea, even till *Mahābhārata* period. It is evident from the *Yātrā parva* of *Mahābhārata* (Chap., verse 3)

kāmyakaṁ nāma dadṛśurvanaṁ munijana priyam
samudragā mahāvegā yamunā yatra pāṇḍva.

Its eastern flow in the form of a small stream also came into being by the time of *Mahābhārata,* as is evident from the story of Vasudeva's wading across the river Yamunā on the day of Kṛṣṇa janmāṣṭamī in August to carry the Kṛṣṇa to the safe custody of Yaśodharā in Nanda Gāon. Its southern stream had already started drying up at KU. except Pṛthudak where Yudhiṣṭhira performed last rite of his relatives.

After *Mahābhārata* Period the southern stream of Yamunā dried up completely. The excavation in the area of Pṛthudak and others has made it clear that there were number of cities which were neither destroyed by floods nor by human attacks. Kālibaṅgā is one of such cities which was abandoned during 1800-1700 BC. or 1300-1400 Kali era as shown by Radio carbon dating of the material recovered from that city. Below the Harappan citadel of Kalibangan are the remains of pre-Harappan township which were destroyed around 2450 and 2300 BC. i.e. 650 and 800 Kali era. This indicates that above cities were abandoned due to shifting of Yamunā's waters from its south-ward flowing mainstream to the eastward flowing sub-stream (named Kālindī) which came into being already in *Mahābhārata* period i.e. around 45 years before Kali era.

Geographic and sedimentological evidences are also now available to make it clear that southward mainstream and eastward flowing sub-stream were deriving water from one and the same channel.

Thus with the diversion of Yamunā from south to east, its southern track went dry and became as a rain-fed river deriving almost all of its waters from Ādibadri of Sirmaur hills of Ambala district. This diversion caused the absence of the Sarasvatī and seeing this absence, as is had from *Purāṇas*, Mārkaṇḍeya Muni did some efforts to feed its old track with water deriving from a channel (known after his name as Mārkaṇḍā) issuing from Ādibadri of Sirmaur hills

Mārkaṇḍeyena muninā saṁtaptaṁ paramaṁ tapaḥ
Tatra tatra samāyātā Plakṣajātā Sarasvatī
Sā sabhājya stutā tena muninā dhārmikena ha
Saraḥsannihitaṁ plāvya paścimena prasthitā diśam

So Mārkaṇḍā Sarasvatī is the last Sarasvatī in the chain.

Could The Life-Span Be Prolonged Through Yoga?

Dr. K.R.K. Mohan*

The epic *Rāmāyaṇa* mentions that the emperor *Daśtatha,* the celebrated father of Lord Rama lived for over sixty thousand years. Similarly many other epics and *Purāṇas* refer to many great men who lived for several hundreds and thousands of years.

HISTORICAL EVIDENCE

Coming to the times of recorded history of mankind, the ancient inscriptions found in Babylonia mention that the great Assyrian King Ashur Bani Pal lived for over seven hundred years. Historians believe that the ancient historians wantonly inflated his age to commensurate with his greatness. But why should those inscriptions give false information while the other aspects are believed not to be so?

In the ancient times the entire world was under the inflaence of Hindu wisdom. Hence it is no wonder that kings like Ashurbanipal achieved life-span dilation and lived for the period as mentioned in the inscritions. In fact the King's name itself was a Hindu one. The name Ashura vānipala meaning the king who controls the voice of Asuras (the enemies of Suras) slightly modified as Asurbanipal. In fact in certain Indian Languages the phoneme 'va' has its allophone as 'ba' - hence Vani became Bani.

It will be interasting to note that *Yogī* type of people existed in Europe too in the past, who were referred to as the 'Men of Mystry.' Nicholas St. Germain, Appolonius of Tanya and Parcelks who lived between 1400-1700 A.D. were among such.

THE CASE OF INDIAN YOGIS

India is a land where great yogīs and Siddha puruṣas with mystic powers existed since the beginning of the mankind. Even today we find many such Yogīs. H.H. Jñanananda Swamy of Tapovanam, Bala Yogī of Mummidivaram (A.P.) Ramnanda Yogī who lived in the present century are a few such. The investigations made at the All India Institute of Medical Science, New Delhi on less accomplished Yogīs Under controlled medical conditions revealed the control and slowing down of many metabolic functions. The investigations conducted at the Para Psychology Department of the Andhra Vniversity also were stated to be very revealing.

INVESTIGATIONS IN THE U.S.A. AND JAPAN

The further investigations continued by Dr. Wegner and Dr.

Bagachi of the University of California and Michigan Medical School also supported these findings. The investigations conducted at the university of Tokyo, Japan on some Buddhist Zen monks had revealed a slowing down of the Metabolic functions, a 20% decrease in the body consumption of Oxygen and a capacity to go into a state of thorough relaxation, when tested with an Electroencephalograph.

The research at Harward Medical School and the University of California had shown in the Yogīs, under meditation, an increase in the skin resistance indication of condition tending towards anxiety-free state, rapid decline in the blood-lactate level which indicated a slowing down of anaerobic metabolic state, decrease in respiration and marked changes in the brain waves signifying a relaxed state.

INVESTIGATIONS AT NASA

With the advent of medical sciences relating to the space where zero gravity exists, the mystery surrounding the phenomenon of dilation of life-span is slowly disappearing. The investigations carried out the NASA. on life sciences revealed the existence of many species in the animal kingdom, capable of life-span dilation through regular periodic hybernation. To comply with the demands of the space travel, many experiments were made not only to understand the phenomena but also to apply them for long duration space travel where the phenomena but also to apply them for long duration space travel where the astronaut would be subjected to a state of 'suspended' or 'slowed down' metabolic state during the major course of the journey. It is an observed fact that in the polar regions many organisms undergo hibernation to combat the bad winter. It has also been observed that certain creatures undergo encystment to tide over the uncongenial conditions. During this period of hybernation of encystment their entire life activity is suspended. Even the life in the seeds of the plants is in a dormant state. It comes into activity only when circumstances suit.

THE RIDDLE OF THE EGYPTIAN MUMMIES AND PYRAMIDS

During the Year 1963, a team of life scientists at the University of Michigan announced their findings that the skin cells of an Egyptian Mummy were capable of regeneration of living. Since then several theories have been advanced as to whether the extremely well preserved bodies (Mummies) which lived in Egypt over five thousand years ago are but the result of life-span dilation phenomenon which could account for the need of enormous mountain-like Pyramids to preserve them. In fact the ancient Egyptians believed that the Mummies would come back to life after sometime. Hence they kept everything ready including food in the Pyramids.

The observations at the Harward Medical School which revealed an increase in the skin resistance in the Yogīs under meditation, tallies with the findings of the skin cells capable of living in the Mummies. Could these Mummies be Yogīs under meditation in Smādhi?

NEW THEORY IN ARCHAEOLOGY

In fact the recent theory in archaeology suggests that visitors from outer-space landed in Egypt in pre-historic times and constructed the Pyramids. The theory is based on the evidence such as paintings of Spaceships, helemeted creatures, curious coincidences in the matching of the Pyramids' dimensions with the distances between the planets and the Sun and still inexplicable engineering feats involved in their construction.

In the light of the above findings we need not wonder if on some day, the Egyptian mummies come back in life to spend the balance of their life. They may ultimately turn out to be men relating to the men of some other planet, who may reveal to us the mysteries of the Universe.

HOW DILATION OF LIFE-SPAN IS POSSIBLE

From the fore-gone paras it could be seen that the phenomenon of life-span dilation is a reality. It is in existence in certain animals and all seeds. It was widely prevalent among the ancient Yogīs and Siddhas of India. Even now it is being practiced by some Yogīs in India. It was also achieved by some great men of the West.

Now the question as to how the dilation of life-span is possible arises. This subject was widely dealt with in ancient manuscripts of India. The ancient Siddha manuscript "*Śarīra Maruda Marma*" places the normal life-span of man at 120 years without dilation. The present generation with an average life span not exceeding 60 or 70 years views even this figure with scepticism. Then what about the figures which place the life span in hundreds and thousands of years? Is the life-span of men sliding down?

The Yogīs control all their metabolic activity by regulating their respiration. A normal man has 15 breaths per minute i.e. 21600 breaths per day. The entire metabolic activity of men, for that matter of all living beings is dependent on respiration only. Because of respiration, Oxygen alongwith the air enters the body and combines with the blood. In the chemical reaction that takes place, energy is released in the form of body heat, which maintains the entire metabolic activity. In the process, the erosion of the body cells also takes place which naturally affects the life-span. If the respiratory activity is reduced, the erosion of the body also gets reduced in the other words the life-span is prolonged if the respiration is reduced. If the respiration is completely stopped it amounts to hybernation.

The Indian Yogīs who knew the secrets of the respiration practised '*Prāṇayāma*' and controlled their respiration by considerable reduction in the number of breaths. That meant they were maintining their bodies by spending less energy, resulting in the dilation of their life-span.

THE MYSTERIES OF RESPIRATION

Even though there are two nostrils inhaling or exhaling never takes place simultaneously, through both the nostrils. If air goes in through the right nostril it comes out of the left nostril and vice versa. By pressing in the arm-pit, the direction of the air passage could be changed. It pressure is applied in the right arm-pit exhaling takes place from the right nostril. It the direction is to be changed to the left nostril, pressure is to be applied in the left arm-pit and vice versa. In fact the direction of inhaling and exhaling automatically changes periodically many times a day depending upon one's constitution. This aspect of respiration is supposed to be not known to the allopathic school of medicine.

Much research is to be done in the process of automatic changing of the direction of the air passage through the nostrils and as to why pressure in the arm-pit changes the directions and the resultant changes that take place in the body.

The ancient Indian Siddhas and Yogīs who knew the secrets of the body used to control their respiration and acquire many mystic powers. While meditating they kept a '*Yoga Dannḍa*' (a curved stick) under their arm-pit to apply pressure. They changed the stick to the other arm-pit whenever they wished to change the direction of the air passage.

THREE STAGES OF THE BODY LOSING CONSCIOUSNESS

There are three stages of the body losing consciousness. The mildest of these stages is the sleep. During this stage the life activity of the body is noticed by the respiration only. The normal five senses are lost and brought down to the minimum.

The second stage when the body loses consciousness completely is the Death. During this stage all the metabolic activities of the body— all physiological functions are completely lost for ever.

The third stage of the body losing consciousness is *'Samādhi'* — a Yogic feat practiced by the Hindu Yogis and Siddhas. They suspend all the body activities including heart-beat. It is virtually physiological death. But the body won't get putrified even if it is kept deep below the Earth. It comes back to life after hours or days as per the will of Yogī, who sets the time to regain consciousness at his will, just as a time is set for exploding a time bomb or an alarm clock.

The state of *Samādhi* is akin to mummification of the ancient Egyptians. But while the latter is achieved with the help of certain herbs and ingredients, the former is devoid of any such thing it is pure hybernation. In all such cases the life span is naturally dilated.

CONCLUSION

From the discussion made supra it is evident that dilation of life-span is possible by the reduction of the number of breaths. The experiments conducted at various institutes confirm this fact that respiration enhances the life-span. Perhaps it might be the reason for the starved people to live more as they consume less food and the oracious eater to die early as they consume more. In the light of these findings it is not difficult to understand that accomplished Yogīs and Siddhas through their years of concentration backed by their faith in the Supreme Being achieved life-span dialation.

Now that the process of hibernation becoming a necessity for long distance and duration space travel much research is to be done in this field. Mastery over many mysterious phenomena in the field of life sciences will unravel the greatness of Hindu Yogīs in achieving life span dilation without the modern electronic or medical gadgetry.

** 11-19, P & T Colony, Dilsukhnagar, Hyderabad - 500060*

Treatment of ENT Disorders in The Vedas

Dr. Ravi Prakash Arya

The Vedas are the record of great advance made first ever by humanity. Vedic scholars were the masters of the vision. They visualized and unraveled the secrets of the spiritual, astronomical and physical world. These secrets were encoded in the Vedas, which were further elaborately exposed in the Upvedas, Brahmaṇas, Āraṇyakas, Upaniṣads and Sūtra literature. Āyurvedic science being one such elaboration and exposition in the spiritual world on human anatomy, medicine and surgery. Āyurveda is classified among Upvedas and regarded as an offshoot of Atharvaveda. Āyurveda is also called as Pañcama Veda. The origin of Āyurveda as a separate school is pointer to the awareness and keenness of Vedic people towards the study of human anatomy, various mental and physical disorders and their best possible remedies. As such ENT. disorders were also studied and examined minutely and their proper panacea was also sought accordingly.

ENT. disorders were studied under Ūrdhvajakrugata Roga heading. Ūrdhvajakrugata Rogas not only involved ear, nose and throat disorders, but the disorders of eyes and head were also included therein. There was in fact, a tradition to read ear, nose and throat together with eyes. e.g.

MukhanāsikākṣIśrotrāṇī (*Praśnopaniṣad*, 35, quoted by UVK, p 489)

However, at certain places nose was also quoted alongwith ear only, e.g.

śravaṇe nāsike na gndhāya (*Nyās Up.* 5, quoted by UVK, p 489)

Sometimes we also find nose coupled with throat only.

Mukha nāsikābhyām (*Garbhopaniṣad,* 3, quoted by UVK, p 489)

The aforementioned observations and statements only mean that eyes, ears, nose, throat and head are the parts of the one and the same channel and each one interconnected with a common track. It is through nose and mouth that not only all these organs, rather say the whole internal system of our body gets exposed to the elements existing in the external environment through the process of inhalation and intake of food and water. When the air is inhaled or food or water is taken in, the nose, throat, ear and eyes develop all likely-hood and possibilities of being infected first by infectious bacteria or vicious germs born in by air we inhale or food and water we intake.

According to the Vedic seers it is only the nostrils that support the atmospheric air in the form of respiration.

Vāyu prāṇo bhūtvā nāsike prāviśat (*Ait. Up.* 2.4)

Since atmospheric air is the part of ultimate respiration or *Prāṇa* (*prāṇād vāyurajāyat* -(RV. 10.90.3), the atmospheric air having entered into nostrils transforms again into *Prāṇa* (inspiration), and *Apāna* (expiration)

These *Prāṇa* and *Apāna* were medically termed as *Aśvinau* with nose as their place of origin.

Nāsikā prabhavau babhūvaturiti vā (*Nir.* 6.3.13)

Similarly, the duet of *Indra* (biological charge) and *Agni* (temperature) of our body were also called as *Aśvinau* with their placement in the mouth.

Mukhādindrścāgniśca (*RV.* 10.90.13)

Aśvinau were considered as great physicians, as is known to all. Since almost all the diseases, a s already pointed out above, are caused either by air-born bacteria or germs inhaled by us through nose or food and water had by us through our mouth, there perfect remedy or cure was also found by Vedists through the ways of (1) inhalation and (2) oral intake of medicinal herbs.

Inhalation consists of (1) *Prāṇāyāma* or the regulation of *Prāṇa* (inhalation) and *Apāna* (exhalation) breathings and maximum exposure of the body to the open air, which is considered to be pure and free from bacteria and germs. This treatment was known as *Aśvinī cikitsā. Ātharvaṇa* seer (AV. 7.53.1) lays down the significance of this treatment for escaping death.

Pratyuhatāmaśvinā mêtymasmaddevānāmsgre bhiṣajā ścibhiḥ

(II) ***Homa*** **or** ***Yajña*** **therapy:** According to the principles of *Yajña* (*Vedic Meteorology*, 1995) a solid or liquid substance when treated with fire gets diffused in the air expanding thousand times in volume in its gaseous state. Inhalation of auṣadhis thus diffused in the air produced effect at large scale similar to the one taken orally. According to the Yajurvedic seer, all disorders including the ones of respiratory system, throat, ears, eyes, etc. can better be cured through *Yajña.*

prāṇaśca me apānaśca me vyānaśca me asuśca me cittam ca me
ādhiśca me dakṣaśca me balam ca me yajñena kalpantām.

Vedic seers also identified such herbs as could purify atmosphere from bacteria and germs when offered to *Yajñīya* fire as an oblation.

(III) *Nasya* therapy: To cure *Ūrdhvajakrugata Rogas* or diseases of nose, ear, throat, eyes and head, *Nasya* therapy, consisted in the intake of certain medicated powdered herbs, liquids, or aromas through nose, was used. The present day Aroma therapy is also an offshoot of ancient *Nasya* therapy.

Oral intake consisted in (I) administration of herbs orally and (II) hydrotherapy.

Apart from above cited two methods; a de-conditioning method was also applied. As per this, patient's body is de-conditioned through Haṭhyogic and Naturopathic exercises, so that it can regain its lost resistance power to fight with all sorts of infectious bacteria or germs and allergens causing various disorders and diseases. These Haṭhayogic and Naturopathic exercises consisted of (I) *Pañcakarma* treatment (e.g. sweating, vomiting, loose-motions, fats' treatment, massage or acupressure), soil therapy, *Śaṅkhaprakṣālana* (the exercise to flush clean the intestinal track), *Dhotī* (exercise to cleanse the upper part of the intestines and vocal cord with the help of cloth made specially for this purpose), *Bastī* (exercise to wash clean the lower portion of intestinal track), *Netī* (exercise to cleanse nostrils with the help of thread or cathedral), *Jalanetī* (cleansing of the nostrils with the help of luck warm water. The latter two exercises directly concern the ENT disorders.

In addition to the above, a special drilling therapy and let blood therapy was also used. For instance to cure certain diseases of ears and nose drilling of ears and nose was done. This was termed as *Nāsāvedha* and *Karṇavedha*. This therapy is precursor to the modern day Chinese acupuncture therapy.

REFERENCES

1. *Ṛgveda* (*RV.*) : Ed. Dr. Ravi Prakash Arya, Parimal Publications, Delhi, 1997
2. *Yajurveda* (*VS.*) : Ed. Dr. Ravi Prakash Arya, Parimal Publications, Delhi, 1997
3. *Vedic Meteorology* : Dr. Ravi Prakash Arya, Parimal Publications, Delhi, 1997
4. *Atharvaveda* (*AV.*) : Devi Chand, Munshiram Manoharlal, Delhi,1982.
5. *Upaniṣad Vākya Koṣa* (*UVK.*) : Col. G. A. Jacob, Motilal Banarssidas, Delhi, 1963.
6. *Nirukta* (*Nir.*) : Brhma Muni Privrajaka, arya Sahitya Mandala, Ajmer, 1966.
7. *Aitareya Upaniṣad* (*Ait.Up.*) : Dr. E. Roer, Bibleotheca Indica, Calcutta, 1850,

Concept of Psychological Disorders and Their Cure in Patañjali's Yoga Darśana

Dr. Surinder Kaur*

From the beginning of human society, people have suffered from sadness, mysterious symptoms, and bizarre behaviour. The physicians and healers on various parts of the Globe tried to classify these disorders, explain them, and cure them. Here it would be interesting to know as to how Patañjali, one of the dominating figure in the Indian wisdom, viewed these disorders. He associated them with the mental process of an Individual, which, according to him could be described as either comfortable / painless or uncomfortable / painful.

Vṛttayaḥ pañcatayyaḥ kliṣṭā 'kliṭāḥ [Yogadarśana (YD) 1.5]

In fact, comfortable processes were described by him as the normal mental functions and uncomfortable processes as the malfunctions.

Uncomfortable or painful processes or malfunctions are further divided into five types, viz. *avidyā.* being first one followed by *asmitā, rāga, dveṣa* and *abhiniveśa.*

Avidyā is said to be the root cause of all others that follow it e.g. *asmitā, rāga, daveṣa and abhiniveśa.*

avidyā kṣetraṁuttreṣāṁ prasuptatanu vicchannodārāṇām (YD. 2.4)

Avidyā, according to Patañjli is the sense of permanence in transient, of purity in the impure, of pleasure in displeasure of consciousness in unconscious.

anityā śuciduḥkhānātmasu nityaśuci sukhātmakhyātiravidyā (YD. 2.5)

Thus the malfunctioning of mental process is identified by Patañjali in utter confusion of opposites, of good and bad, truth and falsehood, pure and impure, permanent and temporary, conscious and unconscious, the essential and existential, etc. It is a confusion as much as a positive and negative, light and dark.

Avidyā causes possessiveness (vfLerk) to follow. *Asmitā* is thoroughly possessive state of mind. An individual becomes highly possessive. Patañjali describes this malfunction as follows :

dṛgdarśanaśaktyorekātmatevāsmitā (*YD.* 2.6)

Asmitā is the blending of an individual together with the objects.

Asmitā (possessiveness) further gives rise to attachment. The individual of possessive nature becomes highly attached to the objects. This attachment is a pleasurable experience. According to *Yoga* philosopher :

sukhānuśayī rāgaḥ (*YD.* 2.7)

'Attachment is the cause of pleasure.'

However, more and more attachment or intense attachment is often said to be the cause of aversion. Aversion or hatred is again very painful.

duḥkhānuśayī dveṣaḥ (*YD.* 2.8)

Last but not the least malfunction of mind is the development of anxiety disorder form of death phobia or desire to live forever *(abhiniveśa).*

svarasvāhī viduṣo'pi tathārūḍho'bhiniveśaḥ (*YD.* 2.9)

According to Patañjali, malfunctioning of mental process can be minimized with the help of some orientation programme popularly known as *Kriyā Yoga.*

samādhi bhāvanārthaḥ kleśtanūkaraṇārthaśca (*YD.* 2.2)

'*Kriya-Yoga* is for bringing about *Samādhi* and minimizing mental malfunctions (*Kleśas*).'

The orientation program or *Kriyā Yoga* is of three-fold. It includes observance of physical discipline counselling and above all faith.

tapaḥsvādhyāyeśvarapraṇidhānāni kriyāYogaḥ (*YD.* 2.1)

Physical discipline is described in terms of *tapaḥ. Tapaḥ* is austerity, hard discipline of body and its habits in order to facilitate change in behavior. It is always observed at the physical level. Counselling is described in terms of *svādhyāyaya.* It is always observed at the mental level. *Svādhyāya* is the self-analysis. It is development of insight into one's own self. It is a short of humanistic approach to therapy. It aims at sharpening the individual's self awareness. *Svādhyāya* is also pointer to the study of good works of literature on life and living, philosophy and ethics. It also includes counselling or *upadeśa* by the person who is well-read and quite senior in experience and khowledge.

Last factor of *Kriyā Yoga* is the faith. Faith is must for any short of treatment you receive for the remedy of your illness. In fact, faith means faith in God. Faith in God helps boost one's confidence and confidence is the key to success.

Thus the above given orientation programme is must for cessation of mental malfunctions (*Kleśa*).

If the mental malfunctions are not ceased, their further aggravation or development takes place into the form of various *cittavikṣepas,* what we call in modern psychological terms as Psychological disorders.

Pātañjali enumerates all these psychological disorders as:

vyādhistyānasaṁśayapramādālasyāviratibhrāntidarśanālabdhbhūmikatvānicittavikṣepāste'ntarāyāḥ (*YD.* 1.30)

First one is *vyādhi* or psychosomatic disorder. Suffering from this disorder an individual complaints sickness without any apparent organic cause.

Second disorder is *styāna* or *ālasya*. *Ālasya or styāna* denotes the disorder of affect like dullness, languor, sloth, drooping state, incompetence, spiritlessness, purposelessness and boredom indicating depression.

Third disorder is *Pramāda*. *Pramāda* denotes carelessness, casualness, drift, deliberate lack of commitment, and earnestness indicating a type of personality disorder.

Fourth one is enumerated as *anavasthitatva*. It may also be taken, in modern psychological terms, as another type of personality disorder. A person suffering from *anavasthitatva* will show his behaviour inconsistent with his words. He doesn't follow his own promises or obligations; he is perfectly willing to deceive and defraud other people; he is always an unstable person or unable to stay in a state.

Fifth one is enumerated as *avirati* which represents the obsessive-compulsive neurotic disorder. A person suffering from this disorder personally becomes indulged both at thought and action level.

Sixth disorder is *bhrāntidarśana* which may be taken for psychotic disorders having confusion, delusion, hallucination etc.

Seventh disorder is counted to be *alabdhabhūmikatva* i.e. inability to recognize objects and their meanings. Hysterical disorders may be represented by *alabdhabhūmikatva*.

Patañjali also reflects light on the psychophysiological symptoms of various types of psychological disorders. According to him, (*duḥkha*) i.e. sadness and *daurmanasya* (frustration), *aṅgameyajatva* (nervousness) and *śvāspraśvāsa* i.e. disturbed breathing are nothing but the psychological symptoms of various types of psychological disorders.

duḥkhadaurmanasyāṅgamejayatvaśvāsapraśvāsāvikṣepasahabhuvaḥ (*YD.* 1.31)

For the information of our readers it may be mentioned that all types of psychological disorders have also been discussed in detail in the *Ayurveda*. The *Ayurveda* deals with all the psychological disorders in the name of *Manasrogas*.

Therapy for Various Psychological Disorders

Thousands of years back, when the rest of world was grappling with these disorders in barbarous ways, Indian wisdom explored this field in depth and recognized the psychological disorders as mental illness, some curable and others hardly curable.

Ayurvedic seers suggested a herbal therapy for various psychological disorders. The present nature of the work does not allow the authors to go into the details thereof.

Here we shall present an account of therapeutic methods suggested by *Yogadarśana* for the treatment of Various psychological disorders.

Patañjali suggests the *Yoga* therapy as the single out treatment for the neurotic disorders discussed above.

Yoga Therapy

The *Yoga* therapy suggested by Pantañjali for the various types of psychological disorders consists of three parts :

1. Training of mind or meditation
2. Relaxation of mind
3. Third one is the training of breath or *prāṇāyām*, or relaxation of body.

Now we shall discuss all three parts of therapy suggested by Patañjali in detail :

1. Training of Mind or Meditation :

All the psychological disorders and physiological disorders affecting mind side by side can be counteracted or treated with the help of meditation.

tatpratiṣedhārtham ekatattvābhyāsaḥ *(YD 1.32)*

'To counteract these disorders, meditation should be practised.'

According to him, meditation is the best way to attain mental relaxation and thereby treating the disorders.

He himself explains the term meditation and the method of meditation. How can one practice meditation of the Principal reality / Supreme self or consciousness?

As to the question of what is meditation he explains meditation as to surrender oneself to almighty. How can one practice meditation of God or surrender to Him? He tells a very simple method. If you want to meditate upon Him or surrender yourself to Him, you should keep Him in your mind each and every second.

tajjapastadartha bhāvanam (*Y.D.* 1.28)

Now one may ask in what form or with what sign one can keep him in mind. Patañjali gives a very simple answer. According to him. *Aum* is the name. He may be referred to by His name *Aum*.

Tasya vācakaḥ praṇavaḥ (*Y.D.* 1.27)

2. Relaxation of mind

Relaxation of mind can be achieved through various

ways :

- First way is mental loosening up. One will feel relaxed if he gets out of mental roots. Patañjali in this regard states as under :

vītarāgaviṣayaṁ vā cittam

'If the mind becomes free from all passions and thoughts, the disorders may be counteracted.'

- Second way of mental relaxation is the undivided concentration or attention. The undivided attention can be generated in many ways. The mind may be directed to attend on one object or thought. According to Patañjali, a habit of mind to attend on one single point / object or thought brings about the relaxation of mind.

viṣayavatī vā pravṛttirutpannā manasaḥ
sthitir nibandhanī (*Y.D.* 1.35)

Mind may be directed, as per Patañjali, apart from attending on an external object to attend on one's own self. Directing concentration / attention on one's own self also leads to the relaxed state of mind.

viśokā vā jyotiṣmati (Y.D. 1.36)

The attention may also be directed to some blissful knowledge revealed in the state of dream or sleep in order to attain relaxation of mind.

svapna nidrā jñānāvalam banaṁ vā. (*Y.D.* 1.38)

Or through concentration on choiced object, one may relax his mind.

yathābhimata dhyānādvā (*Y.D.* 1.39)

According to Patañjali mind may, for the purpose of concentration, be focused on any object ranging from smallest particle to the greatest and highest object.

paramāṇuparama mahttvānto'sya vaśīkāra (*Y.D.* 1.40)

Thus all the three ways, mental loosening up or thoughtlessness of mind, undivided attention or concentration and meditation leads to mental relaxation which causes the end or smoothening of disorders.

3. Training of Breath or Prāṇāyāma

Third part of *Yoga* therapy is relaxation of body. Best way of relaxing the body is observance of prāṇāyāma or regulation and control of breath. There is a saying, sound body keeps a sound mind. If body is not relaxed, how can mind be without distractions and agitations. For relaxation of mind, relaxation of body is must. Disturbed breath is the symptom of mental agitation.

duḥkhadaurmansyāṅgamejayatvaśvāsapraśvāsāvikṣepa sahabhuvaḥ (*Y.D.*1.31)

Regulated and controlled breath will also lead to the peace of mind or mental relaxation.

pracchardana vighārṇābhyāṁ vā prāṇasya (*Y.D.* 1.34)

'By controlled and regulated inhalations and exhalations, mind gets relaxed leading to the counteraction of disorders.'

Patañjali defines *prāṇāyāma* as :

tasminsati śvāsapraśvāsayorgativicchedaḥprāṇāyāmaḥ

'Praṇāyāma is the suspension of inhalation and exhalation by being in the state of some posture.'

According to Vyāsa, a commentator of *Yogadarśana*, when the *āsana* has been perfected, the suspension of inhaling external air, and that of exhaling the internal air i.e. suspension of either way, is *prāṇāyāma.*

Thus there are three types of *prāṇāyāma :* external, internal and stilled.

External *prāṇāyāma* is the suspension of exhalation. Internal *prāṇāyāma* is suspension of inhalation. Stilled *prāṇāyāma* is observed in absence of both the external and internal *prāṇāyāmas,* when breath is neither exhaled nor inhaled, but stilled as it is and where it is, it is called *stambavṛtti* or stilled *prāṇāyāma.*

bāhyābhyantara stambhavṛttirdeśakālasaṁkhyābhiḥ

pridṛṣṭo dīrgha sūkṣamāḥ (*Y.D.*1.50)

Prāṇāyāma is external, internal and stilled. It may be long or subtle when measured in space, time and numbers.

The above mentioned three types of *prāṇāyāma* were later referred to as *recaka,* the external, as *puraka,* the internal and the one as *kumbhaka.*

In fact *kumbhaka* or *stambhavṛtti* is suspension of breath in between inhalation and exhalation.

Patañjali also takes into account the fourth type of *prāṇāyāma.*

The three types of *prāṇāyāmas* discussed above requires inhalation and exhalation. But the fourth one is that which doesn't require inhalation and exhalation.

'Bahyābhyantraviṣayākṣepīcaturthaḥ (*Y.D.* 2.5)

Now about the practice of *prāṇāyāma* one should be in a posture comfortable to him whether sitting, standing or sleeping. If you are in sitting or standing posture, you should also apply two knots. First one is *uḍḍīyāna* knot. *Uḍḍīyāna* knot is applied at the level of neck. Bend your neck to the extent that your chin may touch your chest. This knot is must after you have inhaled the air and suspended it, or you may feel giddiness and headache. Second knot is termed as *Jālandhara bandh. Jālandhara* knot is applied at the level below the naval on the belly. It is always applied after the breath is exhaled and suspended by tightening the belly. The knowledge of these two knots is quite necessary before we go ahead for the practice of *prāṇāyāma.* Thus having perfected in the application of these two knots, one should start with external *prāṇāyāma* first i.e. slowly exhale the air and hold it therewith as long as you can do comfortably. External *prāṇāyāma* is prescribed first, as with the exhalation of breath all thoughts are also removed out of the mind making it completely thoughtless or passionless. This loosening up of mind also helps in its relaxation.

After the external *prāṇāyāma* is completed, one should observe the internal *prāṇāyāma* by inhaling the air very slowly and holding it inside so long as one can do it comfortably. If you start getting tired of holding it inside, breath out slowly. Before you breath it out completely, hold it in between for the third *prāṇāyāma* i.e. stilled one. This practice will lead one to its higher stage of practice, where motion of breath (inhalation or exhalation is hardly perceptible or say breathing and non-breathing becomes almost identical, which is the fourth *prāṇāyāma.* Thus having perfected *prāṇāyāma* upto its fourth stage, one's mind become fit for concentration and attention.

dhāraṇāsu ca yogyatā manasaḥ (Y.D. 2.53)

On the basis of the foregoing discussion it can easily be informed that practice of *prāṇāyāma* is quite helpful in loosening up of mind, undivided attention, concentration or meditation.

Reference

Yoga Darśana of Patañjali : Tr. in Hindi by Prabhu Dayalu, Published by Sri Venkteshvara Steam Press Bombay, Vikram. Saṁvat 2009.

---

* Scientist, Defence Institute of Psychological Research (DIPR) Lucknow road, Timarpur, Delhi.

Book Review

Devayāna Saṁskṛti Kā Vikāsa Aur Veda

Reviewed by

Dr. Ravi Prakash Arya

The book '*Devayāna saṁskṛti kā vikāsa aur Veda*' written in Hindi by Dr. Aruna Kumar published by Śyāma Yaduvaṁśa trust comparising of 22+184 pages (paper back) priced Rs. 120 is to hand for review.

The book is a living example of author's devotion and dedication to Vedic studies. Though professionally associated with the archaeological survey of India, the author Dr. Arun Kumar seems to have evinced a great deal of zeal and enthusiasm towards the survey of Vedic scientific knowledge. I have my all appreciation for the great efforts put in by the enthusiast author for shedding an ample good light on the evolution of Vedic culture in altogether a new perspective. Author's efforts will not only be fruitful in comprehending the cultural and scientific evolution of first ever evolved society of this globe, but also help explore the mysteries of origin of human beings and universe.

Although the present author is not agree with all the propositions formulated by the author with regard to the origin and development of Vedic society and its comprehension of natural and scientific laws.

From the *paurāṇika* records it is crystal clear that life transferred crores of years ago on this earth from some other planet in the universe or our solar system. That planet has been remembered as the *svarloka* or *svargaloka*. Human beings called devas landed on this planet crores of years ago from *svarloka* or *svargaloka* which was plainly some other planet of our solar system. As per present author ,*svarloka* or *svargaloka* represents the planet mars so far as the planets of our solar system are concerned (See *Bhārtīya Kālagaṇanā kā Vaijñanika evaṁ Vaiśvika svarupa*, 1997, Bhartiya Itihasa Samkalana Samiti Delhi). In fact, mars has been enumerated as placed in *dyuloka (RV.* 1.164.10&2.27.8), so also the people landing from mars were known as devas, *Paurānika* memoirs also fortells that life will transfer from earth to Janaloka or planet Jupiter after the earth will be unable to sustain life on it. In view of above observations devas' landing on the earth from mars can not be gainsaid. If it was so, they must have landed upon tibetan plateau which is the highest altitude on the earth and also still known as the highest desert of the world. The same fact has also been confirmed by Swami Dayananda, a great embodiment of Vedic life and thought, while taking for analysis the origin of human beings . As per Dayananda also humans first appeared in Tibet on the earth in the prime stage of their life. Triviṣṭapa or Tibet has also been equated in the *Purāṇas* as *svarga-loka* . The author of the 'Third Eye' also refers to a tradition of the lamas of Tibet in the context of a cave which is recognized among lamas a place where the devas landed first on the earth. The desert like conditions and the highest position must have made the Tibet all the more suitable place for landing by the than extra-terrestrial life.

Recent researches regarding the famous pyramids of egypt have given some clues about the transfer of life to this earth from some other planet in the cosmos. The pictures of spaceships found in the pyramids all point out to the possibility that the persons mummified in pyramids must have landed on the earth from some far or near planet of our solar system or other.

In the light of above facts, the hypothesis of *devāvatarṇa* from *svarga* and the development of a human culture on our planet cannot be ruled out. The memory of their original planet was still there in the minds of extra-terrestrial human-beings, though the very notion of the word continue to change starting from actual planet down to the highest to higher altitudes of the earth, the later generations who were devoid of the science and technology developed by their ancestors found it not only difficult but impossible to reach the same place or *svarga.*

In the third chapter the cosmic science has been discussed briefly but lucidly comparing the modern theories, which are only the repetitions of the age old facts discussed by Vedic *Ṛsis.* The author may also consult here the present author's 'Vedic Meteorology ' which contains a chapter on the origin of universe, as revealed in the Vedas .

For more details regarding *Kālagaṇanā*, the present author's *'Bhāratīya Kālagaṇanā kā vaijñānika and vaiśvika svarupa'* may be referred to.

Here I would like to differ with the view that the man originated in seventh *manvantara*,i.e. *Vaivasvata manvantara* and various names of the *manvantaras* denote various developments in the origin of cosmos. If we go by *Manusmṛti* we'll come to know that various names of *manvantaras* were coined after the names of famous personalities born that time. *Vaivasvata manvantara* itself is known by the name of Vivasvān.

I would also like to differ with one more proposition of the author borrowed by him from Suryakant, (1963:223-37), etc. The *RV.* not only makes the mention of 27 constellations, but it also enumerates them in the names of various deities , For the Vedic names of various constellations see present editor's *'Bhāratīya Kālgaṇana'*(Chp. 15). In addition to the above facts, the author has very beautifully depicted several other sciences developed by Vedic society, such as physiology, anthropology, pharmacology, surgery, prolongation of life, treatment of animals, cloning and fetus transplantation etc. in the fifth chapter.

Made in the USA
Monee, IL
05 November 2026

67374712R00020